SNAKES IN MY GARDEN

Joy Austin

BeaLu Books

Copyright © 2020 by Joy Austin

All rights reserved. No part of this publication may be reproduced in any form or by any electronic or mechanical means, including information storage and retrieval systems, without the express written permission from the publisher, except in the case of brief quotations embodied in critical articles or reviews. For information regarding permission, contact BeaLu Books.

ISBN Hardcover: 978-1-7341065-6-5
ISBN Paperback: 978-1-7333092-8-4

Library of Congress Control Number: 2019952457
Publisher's Cataloging-in-Publication Data is on file with the publisher.

Edited by: Luana K. Mitten
Book cover and interior design by Tara Raymo • creativelytara.com

Printed in the United States of America
October 2019

BeaLu Books
Tampa, Florida

www.BeaLuBooks.com

PHOTO CREDITS: Cover: © K Quinn Ferris; Page 1: © Head_Snake; Page 3: © komkrit Preechachanwate; Page 4: © pixelworlds; Page 5: © tea maeklong, © Illonajalll, © Pakhnyushchy; Page 6: © tea maeklong, © Illonajalll, © Pakhnyushchy, © Eric Isselee; Page 7: © Rusty Dodson; Page 8: © Jay Ondreicka; Page 9: © Pakhnyushchy, © J.W. Taylor; Page 10: © Pakhnyushchy, © Eric Isselee, © clarst5, © fivespots; Page 11: © Patrick K. Campbell; Page 12: © David A Litman; Page 13: © Pakhnyushchy, © clarst5, © Eric Isselee, © fivespots; Page 14: © Eric Isselee, © Illonajalll, © Pakhnyushchy, © Matt Jeppson; Page 15: © Ryan M. Bolton; Page 16: © Bildagentur Zoonar GmbH

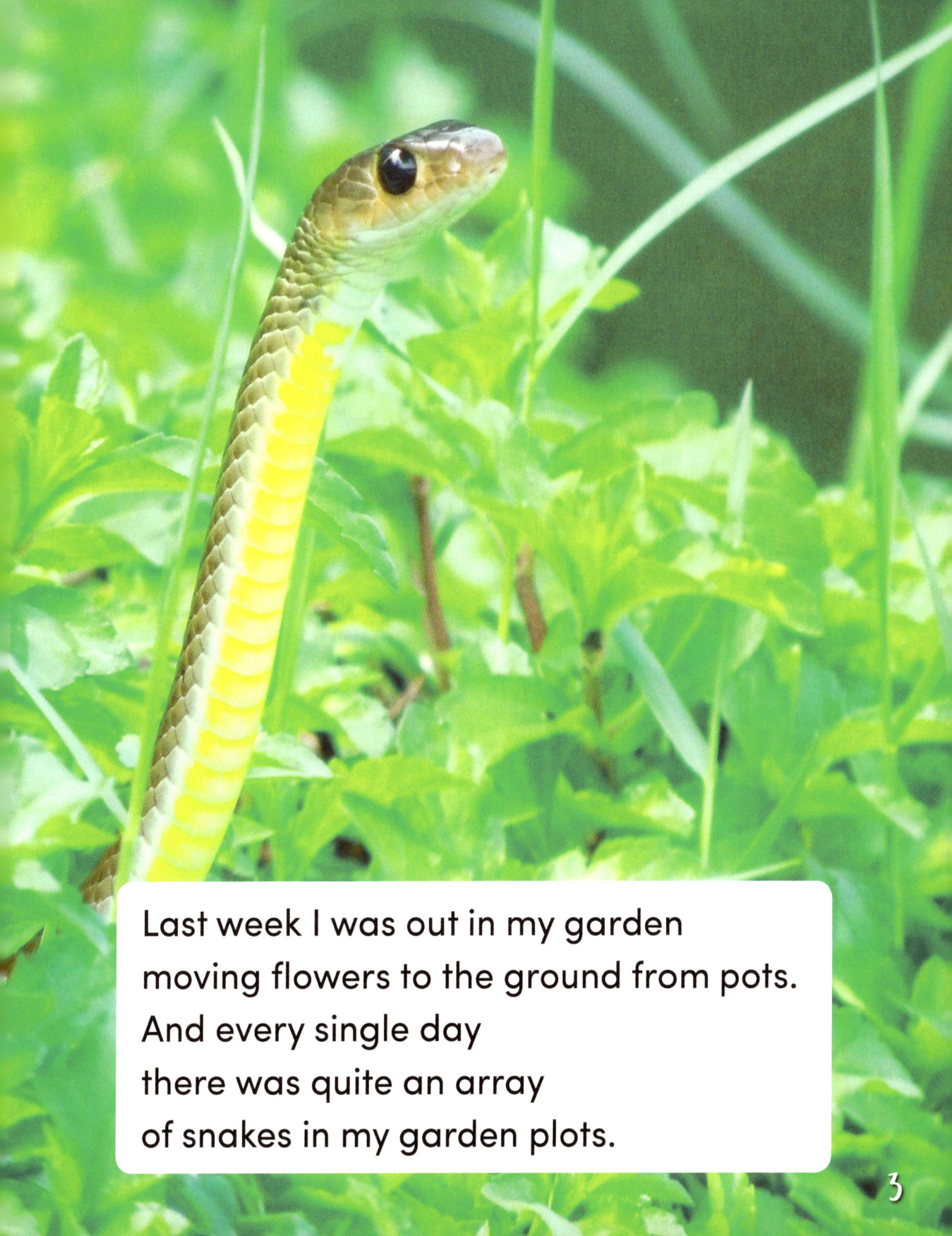

Last week I was out in my garden
moving flowers to the ground from pots.
And every single day
there was quite an array
of snakes in my garden plots.

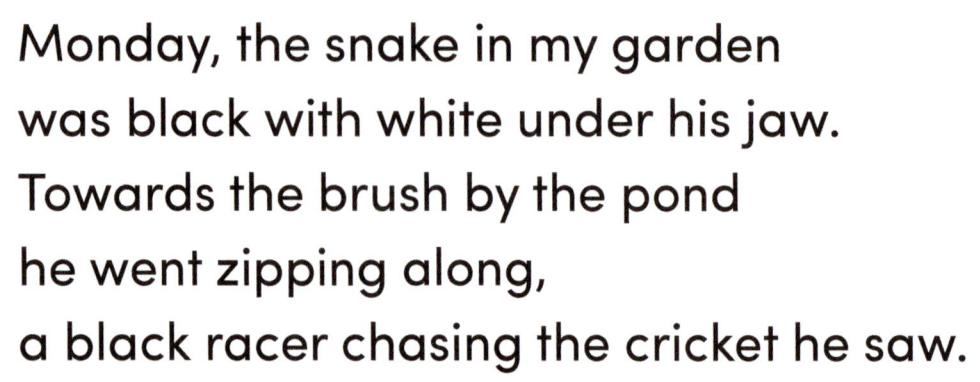

Monday, the snake in my garden
was black with white under his jaw.
Towards the brush by the pond
he went zipping along,
a black racer chasing the cricket he saw.

FACT BOX:

Average Length:
2-4 feet

Food:
insects
lizards
rodents

Fun Fact:
The longest black racer on record was six feet long!

FACT BOX:

Average Length:
2-3 feet

Food:

insects

lizards

rodents

amphibians

Fun Fact:
The ribbon snake is one of the only kinds of snake that does not lay eggs!

Tuesday the snake in my garden
was striped with red, yellow, and brown.
He slithered through the grass
and I waved as he passed,
a ribbon snake chasing frogs down.

Wednesday the snake in my garden
was bright orange with red spots so nice.
He looked so lovely there
I had to just stare
at the corn snake searching for mice.

FACT BOX:

Average Length:
2-4 feet

Food:

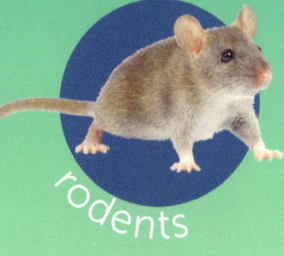
rodents

Fun Fact:
Corn snakes get their name because the markings on their bellies resemble Indian Corn.

FACT BOX:

Average Length:
5-9 feet

Food:

rodents

amphibians

birds

reptiles

Fun Fact:
Indigo snakes eat other snakes, including rattlesnakes and copperheads!

Thursday the snake in my garden
was black from forked tongue to tail.
His dark shining smile
was so meek and so mild,
an indigo snake on a rat's trail.

Friday the snake in my garden
was brown bespeckled with black.
Almost seven feet long
he went slowly along,
a gopher snake rummaging for rats.

FACT BOX:

Average Length:
4-7 feet

Food:

rodents

birds

amphibians

reptiles

Fun Fact:
Gopher snakes may appear aggressive because when threatened they will coil up in a strike position, flatten their heads, and vibrate their tails like a rattlesnake.

FACT BOX:

Average Length:
3-4 feet

Food:

amphibians

lizards

rodents

Fun Fact:
When threatened, hog-nosed snakes will play dead by flipping onto their backs, opening their mouths, and letting their tongues hang out.

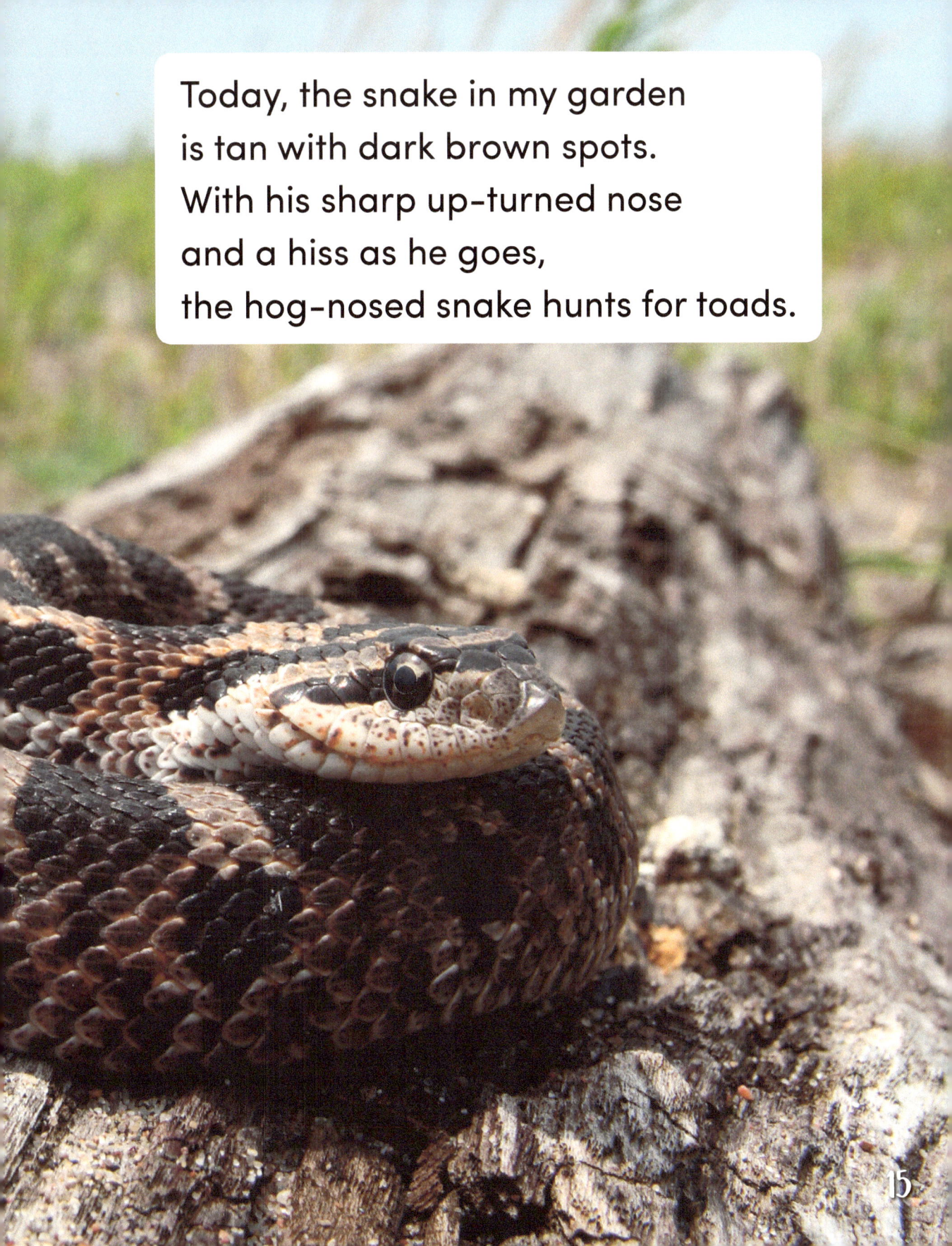

Today, the snake in my garden
is tan with dark brown spots.
With his sharp up-turned nose
and a hiss as he goes,
the hog-nosed snake hunts for toads.

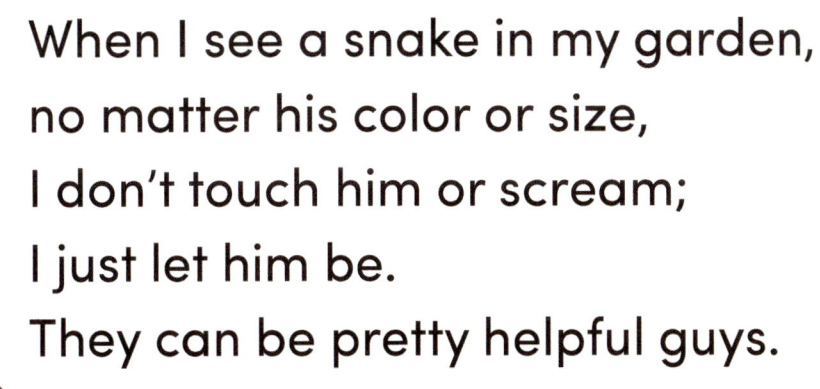

When I see a snake in my garden,
no matter his color or size,
I don't touch him or scream;
I just let him be.
They can be pretty helpful guys.

ABOUT THE AUTHOR:

Joy Austin lives in Tampa, Florida where she works as a high school English teacher. She loves reading and writing poetry. In her free time, Joy enjoys taking road trips to her favorite cities like Charleston, SC. and Boston, MA.

AUTHOR'S NOTE:
I hope you enjoyed learning about some of the common snakes of the southeastern United States. None of the snakes in this book are venomous, but some snakes are. All snakes, venomous or not, may bite when they feel cornered or scared. Just imagine how scared you would be if a giant tried to pick you up! Appreciate them from a distance by giving them plenty of space and never try to handle one even if it's in your very own backyard.

Read more!

Check out the University of Georgia's Herpetology website:

https://srelherp.uga.edu/index.htm